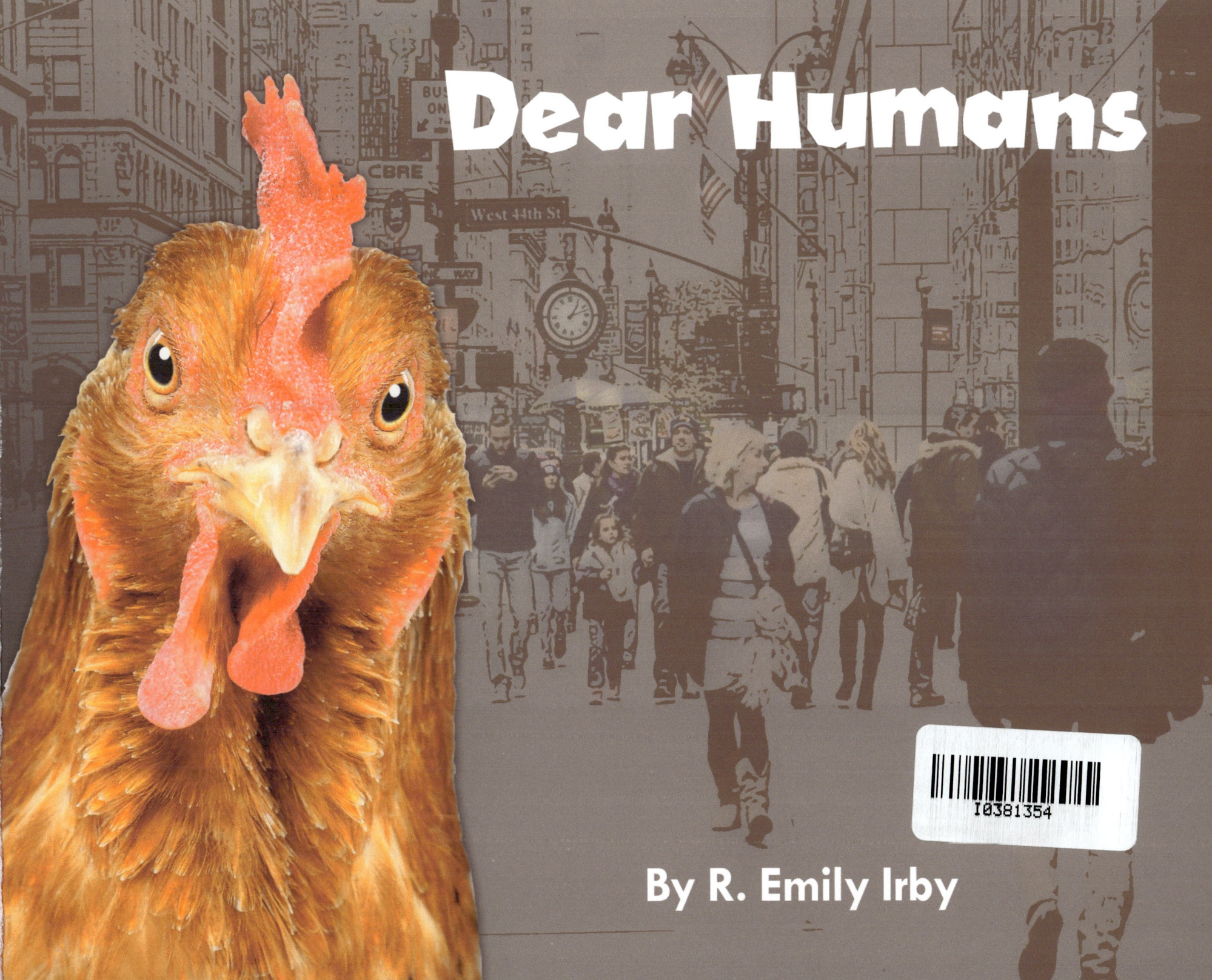

Copyright © 2021 R. Emily Irby

All rights reserved. No part of this publication may be reproduced in any form or by any electronic or mechanical means, including information storage and retrieval systems, without the express written permission from the publisher, except in the case of brief quotations embodied in critical articles or reviews. For information regarding permission, contact BeaLu Books.

ISBN Hardcover: 978-1-7353641-2-4
ISBN Paperback: 978-1-7333092-8-8

Library of Congress Control Number: 2020944523
Publisher's Cataloging-in-Publication Data is on file with the publisher.

Edited by Luana K. Mitten
Book cover and interior design by Ken Silbert

Printed in the United States of America
First Edition
2 7 6 5 1

BeaLu Books
Tampa, Florida

www.BeaLuBooks.com

PHOTO CREDITS:
Cover: (chicken) Sonsedska Yuliia/Shutterstock, (city) littlenySTOCK/Shutterstock; page 3: (city) kmichal/Shutterstock, (chicken) stockphoto mania/Shutterstock; page 4,5: (city) stockelements/Shutterstock; page 5:(chicken) Valentina_S/Shutterstock; page 6: (rooster) monticello/Shutterstock, (feet) AVA Bitter/Shutterstock; page 7: (chicken silhouette, used through book) Alexey Pushkin/Shutterstock, (flying hen) natthawut ngoensanthia/Shutterstock, (flapping hen) pets in frames/Shutterstock, (raccoon) xana69/Shutterstock, (fox) RT Images/ Shutterstock, (cat) LMIMAGES/Shutterstock, (dog) TMArt/Shutterstock; page 8: koonphoto/Shutterstock; page 9: David Tadevosian/Shutterstock; Page 10: Santirat Praeknokkaew/Shutterstock; page 11: (from left to right, all Shutterstock) Valentina_S, Sonsedska Yuliia, TTstudio, Valentina_S, Aksenova Natalya, Oleksandr Lytvynenko (3), yevgeniy11, stockphoto mania, Aksenova Natalya, Olhastock; page 12: Rachael Martin/Shutterstock; page 13: Kentaro Foto/Shutterstock; page 14: Fotokostic/Shutterstock; page 15:(l-r) Rungroj Youbang/Shutterstock, maggee/Shutterstock, Rungroj Youbang/Shutterstock; page 16: (background) Anton Havelaar/Shutterstock, (rooster) oksana2010/Shutterstock; page 17: Anton Havelaar/Shutterstock; page 18: (background) Moonborne/Shutterstock, (inset) PhotoSongserm/Shutterstock; page 19: (background) Anneka/Shutterstock, (inset) rchat/Shutterstock; page 20: (background) Regreto/Shutterstock, (insets top-bottom) saied shahin kiya/Shutterstock, Anneka/Shutterstock, Vinicius Bacarin/Shutterstock; page 21: (background) Tabatha Del Fabbro/Shutterstock, AustinandAugust/Shutterstock; page 22: Aksenova Natalya/Shutterstock; page 23: prapass/Shutterstock; page 24: R. Emily Irby

Dear Humans,

I can't believe it! You keep asking the same question over and over.

"Why did the chicken cross the road?"

Really?
Everyone knows why!

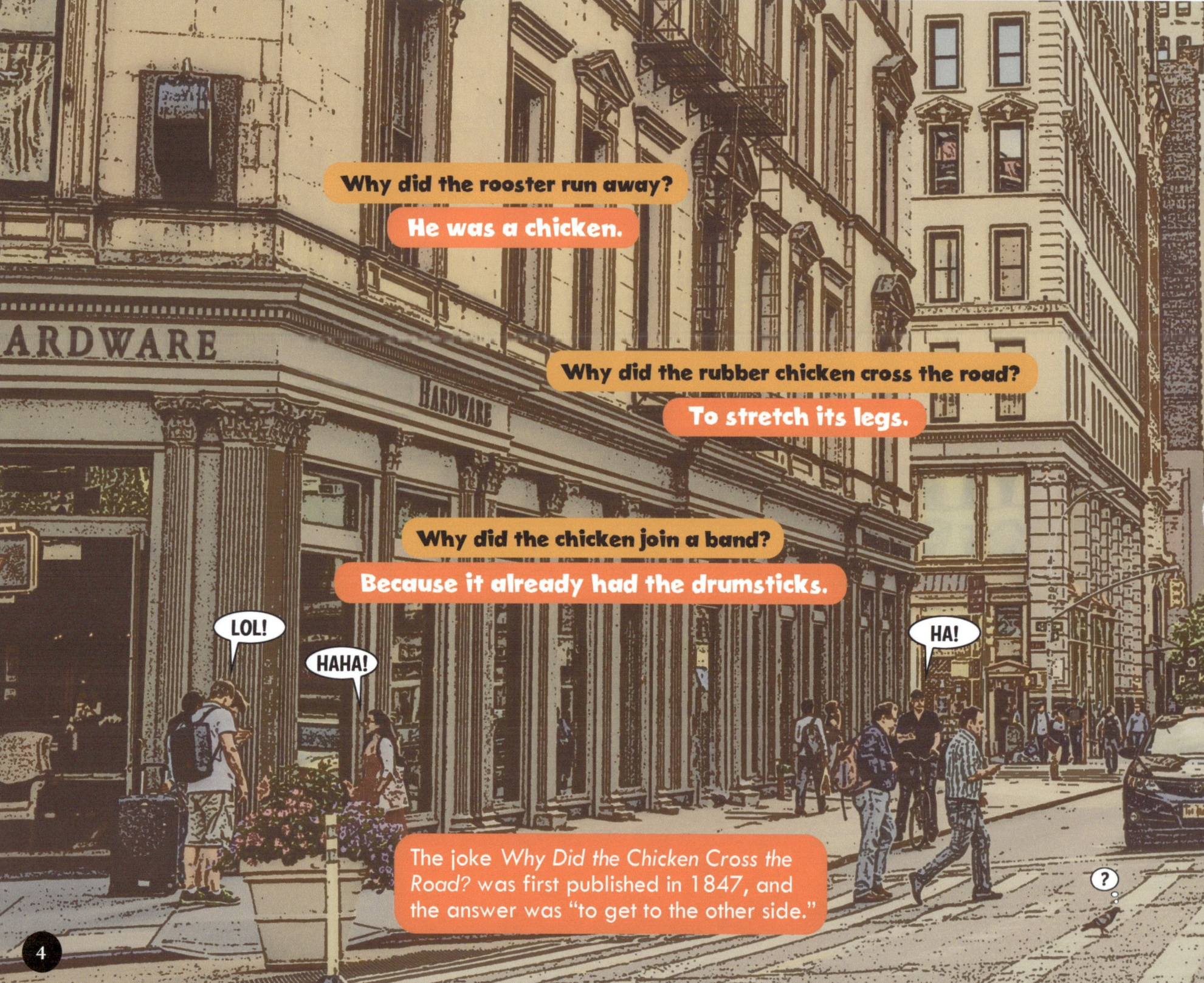

We mighty, courageous chickens are tired of being the center of human jokes. We demand respect!

My name is Gallus, and I am here to set the record straight. I feel that we need to come to an understanding. Chickens and humans are really not that different. Chickens are brave, intelligent, and get hungry just like you!

Gallus gallus domesticus is the scientific name for chickens.

The jokes were bad enough, but now insults? Being called a chicken should be a compliment, not a put-down! If someone calls you a chicken, stand tall and proud because chickens are brave.

It is true we will run and fly—well, flutter—away from danger, but we will also protect ourselves, our homes, and our babies when needed.

This is why roosters and even some hens have a sharp, sturdy spur (talon) on the back of each leg. And believe me, they are pretty effective in fighting off predators and bullies!

CHICKEN KNOWLEDGE Chickens cannot fly long distances like other birds. The farthest recorded flight is 301 feet (98 meters), almost a whole football field. Most chickens only fly far enough to perch on fences, trees, or rooftops.

CHICKEN KNOWLEDGE The average chicken can run up to 9 miles per hour (14km/h) for a short distance. The average human child can run about 10 miles per hour (16km/h) for a short distance.

Chickens are often prey for raccoons, foxes, cats, dogs, and other predators.

You heard me correctly, bullies. Sadly, another thing we have in common with humans is that some of us are not always kind. Once in a while, a rooster or hen thinks they can pick on a weak or sick chicken. If the bullied chicken cannot flee, they need to defend themselves. Both humans and chickens could be kinder and share.

Did you know the sayings "hen pecked" and "pecking order" come from how chickens pick on each other?

CHICKEN KNOWLEDGE

CHICKEN KNOWLEDGE: When the flock is too crowded and food is scarce, stronger chickens may chase off or harm weaker chickens

We are not "dumb clucks," either. Chickens are smart! Why, in some cases, we outsmart humans! Well, human babies at least. A two-day-old chick can understand that objects still exist even when they are no longer visible (seen). It takes your human babies about seven months to figure that out! So, go ahead and call someone a "bird brain"—it is a compliment!

CHICKEN KNOWLEDGE

Chickens have mighty memories as well. They know which chickens are their friends in the flock, and they can be trained to come when called.

CHICKEN KNOWLEDGE

Chickens can recall over 100 different faces, including human faces. Studies have shown that chickens have complex thinking similar to primates.

Just like humans, we get hungry, and that memory of ours comes in handy. It is important to remember where to find the best meals. Chickens need to eat a healthy diet too. We are omnivores, which means we enjoy fruits, seeds, grains, and even meat. There is nothing like a cool cucumber on a hot summer day. On special occasions, we even get to feast on frog legs, just like humans.

CHICKEN KNOWLEDGE: Chickens that live in overcrowded areas or lack protein in their diet will eat eggs — shells and all!

Believe me, we love to discuss our favorite foods and where to find them, in our own language, of course. There is definitely more going on in our walnut- sized brains than you realize.

CHICKEN KNOWLEDGE

Chickens have about 24 different sounds, each with a specific meaning. A short, low repeated sound means "stay close." A short, high-pitched repeated sound means "there is food over here." A low-pitched purr says, "let's stick together."

After all Humans, we have co-existed with your species for nearly 10,000 years. You are outnumbered.
It is time for understanding and respect.

Sincerely,
Gallus
Your friendly neighborhood rooster

CHICKEN KNOWLEDGE It is estimated that there are three chickens for every one person in the world.

FEELING LIKE YOU NEED A FRIEND?

Why don't you try hatching your own brilliant little buddy? Make it two or more, though. Chickens prefer to be part of a flock even if it is a small one! Being a lone chicken is highly stressful.

Hatching eggs is no job for a dumb cluck! It takes patience! A broody hen must sit on her eggs for 28 days until they hatch. During that time, she turns the eggs and keeps them at the perfect temperature and amount of humidity. And believe it or not, she even talks to them. The developing chicks hear her voice and know that she is their mama.

CHICKEN KNOWLEDGE

Before you get fertilized eggs from a farmer or hatchery, you need to have an incubator for the eggs. An incubator does the job that a hen does in nature when she sits on her eggs.

Did you know turning the eggs helps the chicks develop and prevents them from sticking to the shell.

CHICKEN KNOWLEDGE

Instead of taking 28 days to hatch, eggs in an incubator will hatch in 21 days. While your eggs are in the incubator, you must monitor the temperature, mist them with water, and turn them at least three times a day. And just like the hen, it's a good idea to talk to the eggs, so they know you are their mama.

When your chicks are ready to hatch, they will use their egg tooth to break their shell and slowly work their way out. Chicks need to hatch by themselves without any help from you. The egg tooth will fall off shortly after hatching. Chickens hatched in incubators are friendlier to humans because you are the first thing they see and become mama to them.

Congratulations! Your new buddies have hatched and are ready to leave the incubator for a bigger space. A box like a dog crate with high walls and lots of air circulation works well for your chicks' new home. Besides a box, you need a heat lamp to keep the chicks warm, containers for food and water, and newspaper and chopped straw or pine shavings for the bottom of the box. The chickens will live in the box until all their feathers come in.

Even though your chicks have hatched, you still have lots of work to do. It takes work to care for animals. Each day you need to change the straw bedding in the box and give them fresh food and clean water. As your chicks grow, you may have to get a bigger box for them.

If you want your chicks to be friendly buddies, you should hold them and play with them every day. Chickens are smart and like to play games. They love to race and will even try to catch treats like peas or grapes if you toss them in the air.

When your chicks are about six weeks old, they will be old enough to move from the box to a chicken coop outside. To keep your buddies healthy, you need to keep their chicken coop clean, feed them healthy foods and fresh water, and build them a safe area to roam around. Healthy chickens can live for five to ten years and give you plenty of fresh eggs for breakfast.

CHICKEN KNOWLEDGE

Always wash your hands after holding the chicks or chickens. They can carry germs that can make you sick.

KNOW YOUR HENS AND ROOSTERS

First things first, it's hard to tell if your chicks are hens or roosters until they are three to six months old.

HENS

- Hens are called pullets until after their first birthday.

- Hens are female chickens.

- Hens are not quiet, they make lots of different sounds, but they rarely cock-a-doodle-do like roosters.

- Hens' feathers are usually rounded.

- Hens usually have smaller combs and wattles.

- Hens lay eggs for about two years of their lives.

- Hens are called broody hens when they sit on and hatch their eggs. Broody hens will puff up their feathers and get very defensive when they are sitting on their eggs. They may even peck you or make a growling sound. It's best to stay away from a hen and her eggs!

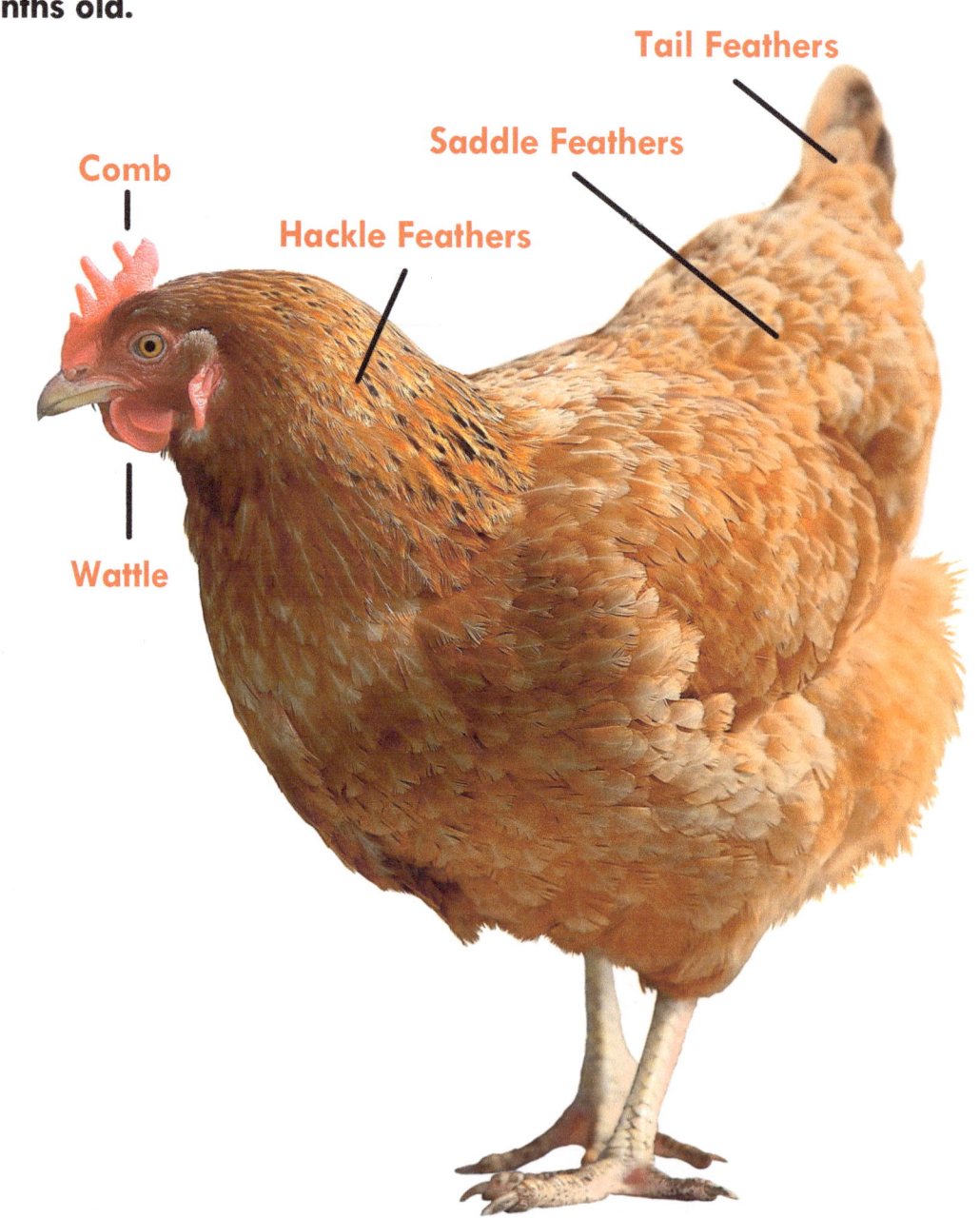

KNOW YOUR HENS AND ROOSTERS

ROOSTERS

- Roosters are male chickens.

- Roosters are called cockerels until after their first birthday.

- Roosters are known for their early morning cock-a-doodle-dos.

- Roosters' feathers are usually pointed.

- Roosters usually have tall, upright combs and larger wattles.

- Roosters usually have colorful feathers. Some of their feathers may even be shimmery. But don't let the shimmery feathers fool you, roosters can be very aggressive and bold. Some even chase humans and dogs. They are courageous and will protect their hens at all costs.

CHICKEN KNOWLEDGE: Did you know the saying "Don't get your hackles up" comes from the way chickens raise their hackle feathers when they are alarmed?

23

ABOUT THE AUTHOR

R. Emily Irby is an intermediate teacher living on the coast of Florida with her husband, son, their two dogs, and a kitty. She loves writing about growing up in a farming community in western Michigan. Her early life and the poems her grandmother wrote to her inspire her stories about animals and family history. Mrs. Irby believes that writing is a creative outlet that can benefit both writer and reader. Her goal is to help students grow to understand, appreciate, and enjoy writing.

The author and her grandmother

"Write to remember, to create, to wonder. Write to express your thoughts, joys, fears, and frustrations. Your pen can be a powerful friend."

—R. Emily Irby

CHICKEN KNOWLEDGE

- The average hen lays about 530 eggs in her lifetime.

- Daylight affects the number of eggs a hen will lay. They need at least 16 hours of daylight to do their best work.

- Hens will stop laying eggs when they molt, feel stressed, do not have a proper diet, and as they age.

- The world's oldest chicken died in 2001. It was sixteen years old.

www.ingramcontent.com/pod-product-compliance
Lightning Source LLC
Chambersburg PA
CBHW041819080526

44587CB00005B/148